11/06

03/2013
3
11/2010

EXTREME ENVIRONMENTAL THREATS™

RADIOACTIVE WASTE

Hidden Dangers

D. D. Kelly

The Rosen Publishing Group, Inc.
New York

Published in 2007 by The Rosen Publishing Group, Inc.
29 East 21st Street, New York, NY 10010

First Edition

Library of Congress Cataloging-in-Publication Data

Kelly, D. D.
Radioactive waste: hidden dangers/by D. D. Kelly.—1st ed.
p. cm.—(Extreme environmental threats)
Includes bibliographical references and index.
ISBN 1-4042-0745-7 (library binding)
1. Radioactive waste disposal in the ground. I. Title. II. Series.
TD898.2.K45 2006
363.72'89—dc22

2006000165

Manufactured in the United States of America

On the cover: Steel drums of low-level radioactive waste fill a trench at this disposal site located at the Hanford Nuclear Reservation in Washington State. **Title page:** These aluminum drums in South Carolina are temporary storage for spent fuel from a research reactor. The nuclear waste will eventually go to an underground waste storage site at Yucca Mountain, Nevada.

Contents

Introduction

This sign warns people to stay away from a landfill containing soil, building materials, and debris that were contaminated with radioactive waste.

Just over a century ago, nuclear power was impossible for people to imagine. Since then, however, physicists have discovered how to split atoms, releasing massive amounts of energy, and engineers have figured out how to channel that energy to make electricity. Today, millions of people in the United States get their electricity from nuclear power. Nuclear power is often called "clean" power because nuclear power plants do not produce the smoke, smog, and global-warming pollution that come from coal and oil power plants.

Nuclear power does, however, produce radioactive waste. This waste can last for thousands of years and can potentially hurt people if they are exposed to it. In the early years of nuclear power, before the risks were understood, nuclear waste was buried or stored in ways that allowed contamination to spread to the environment. Now, government agencies and power companies are tackling the challenges of cleaning up old waste sites and making sure new radioactive waste is properly handled.

Most of us will never see nuclear waste up close. It will be sealed away in metal drums or thick concrete containers. It may be stored in high-security facilities with guards and fences, or it might be buried deep in the ground. But if we could look inside those drums and casks, or travel in the underground vaults, what would we see? We might see a shoe, a glove, or a shirt that was contaminated with radiation and became nuclear waste. We might see a glass block or concrete or metal that contains material that was once part of a nuclear weapon. We might see long metal rods, which hold the fuel once used in nuclear power plants. Nuclear waste can come in many different shapes, but it must all be carefully stored and transported to prevent any radiation from escaping.

1 DISCOVERING THE POWER OF FISSION

This image of a devastated Hiroshima, taken some time after the atom bomb was dropped, shows the immense power of nuclear fission.

The first atomic bomb was dropped in 1945. The story of nuclear power and nuclear waste began, however, in 1896, in a chemistry lab in Paris. At that time, many scientists were searching for new chemical elements and working to identify their properties. Three scientists, Henri Becquerel and Marie and Pierre Curie, had discovered that pitchblende, a black rock, could emit energy rays. They worked to figure out which chemical elements made up pitchblende and which of those elements gave off the remarkable rays.

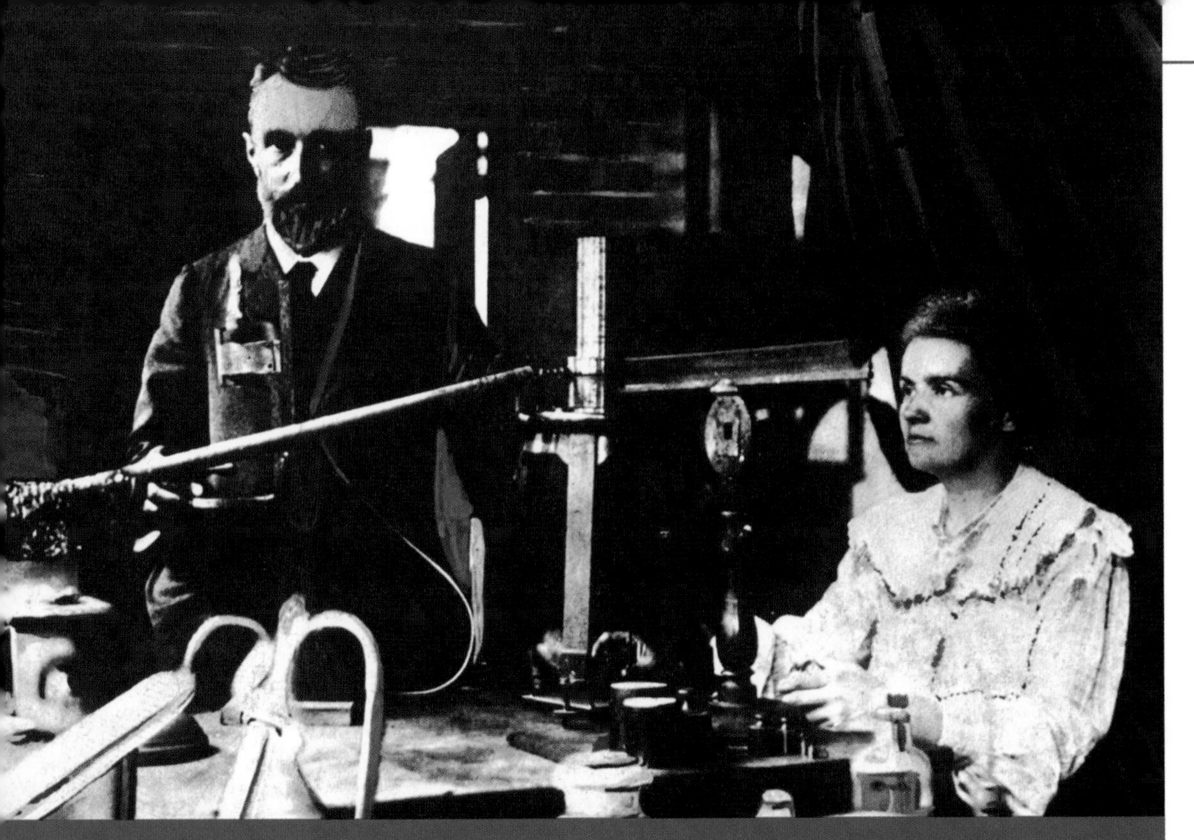

The French scientists Pierre and Marie Curie, shown here at work in their laboratory, found that radiation was emitted from elements in pitchblende. Their discovery of nuclear radioactivity helped scientists better understand how atoms behave. It also led to many advances in technology, including medical treatments, nuclear weapons, and power plants that make electricity.

The Curies discovered two new elements in the pitchblende. These were polonium, which they named for Marie's homeland, Poland, and radium, named for rays of energy given off by the element. Marie made up the term "radioactivity" to describe the special ability these elements had to emit rays that could go through paper, wood, or even metal. The pitchblende also contained uranium, an element that had been discovered previously but was not known to be radioactive. Becquerel and the Curies showed that uranium was radioactive, and that some of the radiation had a positive electric charge, some was negative, and some neutral.

BUILDING THE FISSION BOMB

By the time Marie Curie died in 1934, a few scientists had begun to suspect that if a large number of uranium atoms could be split all at once, immense energy could be released, producing a bomb thousands of times more powerful than any bomb ever before created. In the late 1930s, as Adolf Hitler took control in Germany and began to invade other countries, physicist Albert Einstein realized the threat that Hitler posed to the world. In 1939, he wrote a letter to President Franklin D. Roosevelt explaining the potential danger of a fission bomb.

Hitler took control of uranium mines in eastern Europe in the 1940s, and President Roosevelt worried that Hitler's scientists would find a way to tap the energy of uranium fission. In a top-secret race to make a bomb before Nazi Germany, Roosevelt hired over 100,000 scientists and other specialists to work at labs located in different parts of the United States. By July 1945, these scientists had succeeded. And although Germany had finally surrendered two months before, the United States was still locked in a war with Japan.

On July 16, 1945, the scientists secretly tested their work. A tower in the New Mexico desert was loaded

Atoms and Chemical Elements

Atoms are made up of three basic types of particles: protons and neutrons, which cluster at the center of the atom and together make up the nucleus; and electrons, which orbit around the nucleus.

A chemical element is made of only one kind of atom. Everything we see or touch in the world is made up of atoms of chemical elements. These elements bond together in different ways to form molecules. For example, water is a molecule made of two elements, hydrogen and oxygen. A water molecule has two atoms of hydrogen (H) and one atom of oxygen (O), so chemists named it H_2O. Atoms, when combined, can be dramatically different from their original state. For example, at room temperature, hydrogen and oxygen are gases. When they are combined to become water, however, they form a liquid at room temperature.

This image of an atom shows protons and neutrons (in yellow and blue) tightly clustered in the nucleus. Electrons (white) orbit the nucleus. If the protons and neutrons are split by a chemical reaction, a burst of energy will be released.

with over 100 tons (91 metric tons) of explosives that were woven with tubes of plutonium, a radioactive element.

At 5:29 AM, the tower exploded with an intense flash. A fireball, in the shape of a mushroom, boiled up miles high. It lit the surrounding mountains brighter than the midday sunlight. From bunkers about 6 miles (10 kilometers) away, scientists watched, peering through dark glass, feeling the heat on their skin. It took forty seconds before a strong wind released by the blast hit the bunkers.

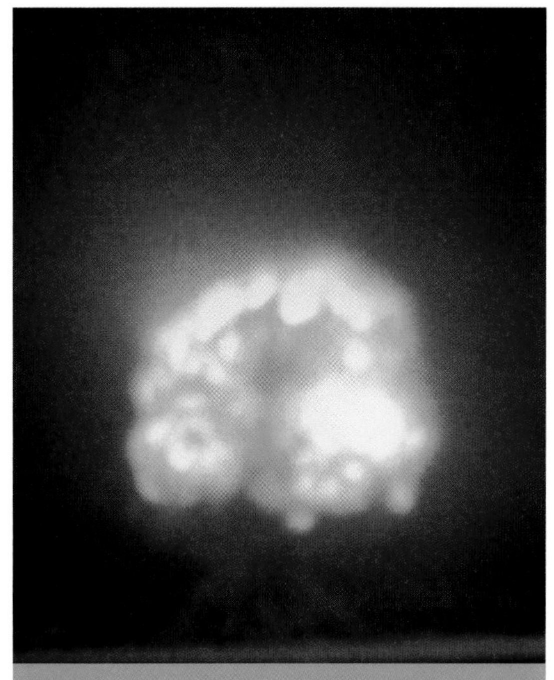

The first atomic bomb, tested at Alamogordo, New Mexico, on July 16, 1945 *(above)*, released harmful radioactivity to the surrounding area. Scientists are still learning about the effects of radioactivity on human health.

In a 1995 article in the *Seattle Times*, Joe McKibben, the physicist who flipped the relay switches that detonated the trial bomb, remembered, "It was a big ball of fire, brilliantly colored and highly turbulent. The color was somewhere between red and purple." The fire could be seen up to 250 miles (415 km) away.

Uranium and its Isotopes

Uranium is a heavy element, with 92 protons and 143 to 146 neutrons in its nucleus. (Compare this to lightweight hydrogen, with a nucleus of only one proton and no neutrons.) Most elements can have different isotopes. An isotope is an element with the same number of protons and electrons but a different number of neutrons. Because neutrons have no charge, they don't change the chemical properties of the element. The isotope of uranium with 92 protons and 143 neutrons is called uranium-235 (92 + 143 = 235). The isotope with 146 neutrons is called uranium-238. Uranium-235 has a special property. Its nucleus is unstable and tends to split (or fission) easily. The unstable atoms break down and give off protons, electrons, neutrons, clusters of particles, or energy. We now know that when uranium atoms undergo fission, eighty different kinds of particles or energy waves may be produced. When neutrons and protons break off from the nucleus of uranium-235, heat is produced and radioactive energy is released.

In the July 1945 test at Trinity, New Mexico, scientists had shown it was possible to compress together enough fissionable material to cause a massive explosion. In a chain reaction that occurred in an instant, the splitting atoms bombarding one another with particles had caused more atoms to split and release still more particles.

Less than a month later, in August 1945, the United States dropped atomic bombs on Hiroshima and Nagasaki, Japan, ending World War II.

2 PROTECTING AGAINST RADIATION

A nuclear waste operator wears protective clothing to keep him safe while he sorts low-level radioactive waste.

It took years for scientists to realize that high doses of radiation could be harmful to humans. Henri Becquerel and the Curies won the Nobel Prize in 1903 for their discoveries, but they did not know that radiation could damage the human body. In 1934, Marie Curie died of leukemia, a cancer of the blood. Doctors now believe it was her work with radiation that caused her disease and death. While most scientists and doctors agree that radiation in large amounts does harm people, there is still much to learn about low doses of radiation. Some

Radioactivity

Some substances are naturally radioactive, such as uranium, and others become radioactive after they are bombarded with neutrons. A radioactive atom is unstable, so it releases particles and bursts of energy. As the atoms in a substance release these particles, the substance is said to be "decaying." Decay creates new elements that are stable and no longer radioactive. Some substances decay in a few minutes, but others take millions of years. Plutonium, for example, decays in 4.5 billion years.

scientists, called epidemiologists, look at populations to see whether people who are exposed to more radiation have a higher rate of cancer. Others, who are able to learn a lot about low dose effects by looking at instances of high dose effects, look at what happened to people in Nagasaki and Hiroshima, Japan, after the United States dropped the atomic bombs that released radiation.

THE BAD LUCK MINE

The U.S. government today requires nuclear waste to be stored and transported with great care so as not to release radiation. In addition, special safety equipment must be given to people who mine uranium ore. But people have not always been so careful. Back in the 1500s, when miners worked in the silver mines of eastern Europe, they found the black pitch-blende rock the Curies later studied. The miners called the rock "pech-blende." "Pech" meant black in German, and it also meant bad luck. They may have called it bad luck because no one would pay for it, and there was more useless pitchblende in the mines than silver. Or, they may have called the rock bad luck because it made them sick. Men who worked in the silver mines died of lung diseases. Some blamed the ore dust, and they wore lace covers over their mouths and noses to filter the dusty air.

Pitchblende, which the Curies studied, is a mineral that contains uranium oxide, a source of uranium. The rock is dense and black or dark green, with a metallic luster and cube-shaped crystals.

A CURE THAT MIGHT MAKE YOU SICK

Years later, after the Curies discovered that pitchblende contained uranium and other radioactive elements such as radium, the mines again became busy. Radium was hailed as a cure for cancer because doctors had discovered that it made tumors shrink. The miners now earned large sums of money for every gram of the black rock they mined. In about 1910, the Marie Curie-Sklodowska Radium Palace was built near the pitchblende mines. Visitors came to drink radioactive water and to breathe radioactive air because they believed it would improve their health. They were probably doing more harm to their bodies than good.

After World War II, doctors in the United States treated cancer with radioactive elements, such as cobalt-60 and cesium-137, gathered from nuclear power reactors. To this day, radiation is one of the therapies used for cancer because radiation in very small doses can kill cancerous tumors. But we are still learning just how much radiation is safe.

STUDYING LARGE GROUPS OF PEOPLE TO LEARN FROM TRAGEDY

To study how radiation affects people, epidemiologists may gather medical records from hospitals, look at death records, and interview people about their health. One such study is under way in the Urals area of Russia. Nuclear weapons are built there at a plant called Mayak, and because of several accidents at the plant, workers and people who live nearby breathed radioactive particles and were exposed to radiation. Farm crops were also exposed, so people ate food contaminated with radioactive material. Scientists will see whether these people are more likely to suffer from cancer and other illnesses than groups of people who didn't have this exposure.

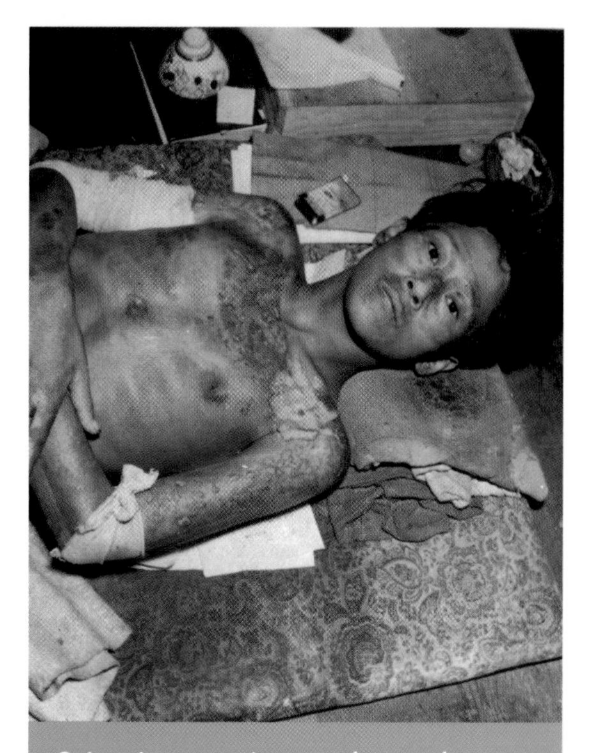

Scientists continue to learn about the effects of radiation by studying groups of people who were exposed to it in large amounts, such as the victims of the Hiroshima and Nagasaki atomic bombings in Japan at the end of World War II.

Some scientists have studied the health of people who survived the atomic bombings of Hiroshima and Nagasaki. When the bombs were dropped, no one anticipated the deadly effects of radiation released by the bomb. Weeks after the explosion in Hiroshima, survivors with no obvious injuries started to develop odd symptoms. Their skin broke out in red spots, their tongues turned black, their hair fell out, and some grew weak and died. Pregnant women gave birth to babies with birth defects. These people were victims of radiation sickness.

Their suffering has helped scientists learn how radiation can harm people.

In 1986, in what is now the country of Ukraine, one of the reactors at the Chernobyl nuclear power plant exploded. The chain reaction in the reactor went out of control, creating a fireball that blew off the reactor's heavy steel and concrete lid. The Chernobyl accident killed more than thirty people instantly and spread radiation about 93,000 square miles (238,063 square km), an area equivalent to the size of England. About fifty rescue workers died immediately after the accident from radiation exposure. Much of Chernobyl's population moved away and has not returned. However, more than 500,000 people who helped with the cleanup have since suffered health problems that may have been due to the accident. Scientists are studying these people to better understand radiation health effects.

PROTECTING AGAINST RADIATION

The people of Nagasaki, Hiroshima, and Chernobyl suffered very high doses of radiation. Small amounts of natural radiation can be found anywhere in the world. Radiation can enter the body through air we breathe or food we eat. Most of the world's radiation comes from a gas in the earth called radon, which can enter homes through basements or in drinking water. The sun also

Many of the health effects of nuclear radioactivity may take years to appear. Scientists and doctors are studying the consequences of radioactivity released in a nuclear accident about twenty years ago in Chernobyl, Ukraine. Radioactive waste produces much lower levels of radioactivity than were released in the 1986 Chernobyl accident *(above)*, but scientists can learn by studying the effects of such accidents.

sends radiation to the earth, and radiation in the form of cosmic rays comes from outer space.

Before people fully understood the risks of radioactive materials, workers who handled nuclear waste were exposed to radiation, and some radioactive waste was improperly handled. Some waste was buried without enough care so that radioactive material seeped into lakes and ponds or traveled to the ocean. For example, some nuclear waste from experiments conducted in

The Geiger Counter

Because radiation is invisible, scientists needed to find a way to detect it. In 1908, several scientists, including Hans Geiger, invented an instrument they called a Geiger counter. It consists of a gas-filled tube with glass at one end and an electrode at the other. Radioactive particles entering the Geiger tube create voltage at the electrode. This causes the counter to beep. The more radioactivity it detects, the faster the Geiger counter beeps. The Geiger counter was one of the first instruments developed to measure radiation, and it is still useful, though there are many other devices now available.

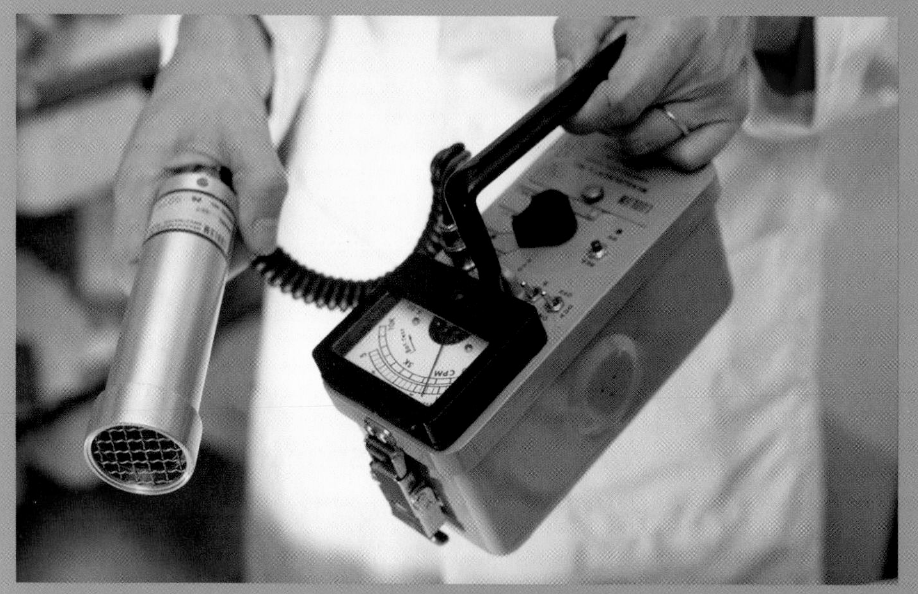

The Geiger counter was one of the first instruments developed to measure radioactivity. Scientists now use a variety of instruments to detect or measure radioactivity. These detectors use different kinds of technologies, including photographic film and ion chambers.

the 1940s was buried at a forest preserve in Palos Hills, Illinois. It was buried in only two feet of soil and covered with a cement cap. Over time, the waste seeped into nearby water. When scientists tested the water from ponds near the waste site, they found high levels of radioactivity. Some radioactive material that was dumped in the ocean has found its way into fish and birds. A January 2003 study, for example, reported that scientists had found radioactivity in Arctic birds' droppings.

Now that there is a better understanding of how radioactivity can negatively affect health, governments have written many laws to protect people and the environment. Workers who mine uranium, build nuclear weapons, make nuclear fuel, or handle the fuel at power plants must wear protective clothing. They wear masks called respirators over their noses and mouths to filter out dust and gas. They also wear head covers, gloves, and coveralls to prevent skin contact with radioactive material. In addition, workers may wear radiation badges to measure the amount of radiation in the area in which they work.

The environment is now also protected from radiation. Ocean dumping is no longer allowed. Nuclear waste must be stored in specially designed casks made of steel and concrete to hold the heat and radiation and keep it from escaping. Transport containers must be able to survive a drop of thirty feet (about ten meters)

These specially designed casks, filled with spent fuel, are being shipped by rail to a research facility in Idaho. The casks, which weigh about ninety tons (eighty-two metric tons) when fully loaded, are designed to shield their contents so no radioactivity can escape. They are also strong enough to withstand a train crash or other accident.

onto a hard surface, and remain undamaged if engulfed in very hot flames for thirty minutes. These scenarios represent the worst possible accident conditions that could occur on road or rail. Trucks and trains must also follow strict rules and procedures when they move nuclear waste across the country. Strict rules must also be followed when the waste is buried.

3 MAKING POWER WITH NUCLEAR FISSION

Large cooling towers, like the ones shown here, release water vapor from nuclear power plants.

With the end of World War II, the United States government decided to find peaceful uses for nuclear fission, such as making electricity. In the 1950s, the U.S. Atomic Energy Commission promoted nuclear power as a source of cheap and never-ending energy for lighting homes and running factories. President Dwight Eisenhower, in 1953, applauded the creation of Atoms for Peace, an organization of scientists committed to finding nonviolent uses for atomic energy. In a 1954 speech, he promised power "too cheap to meter."

In the 1950s, there was such enthusiasm for nuclear power that companies used the atom in their advertising, and children's toys and games boasted atomic themes. In a board game called Uranium Rush, for example, children used a toy Geiger counter to find uranium ore. Children collected "atomic" rings and trucks with "atomic" brakes, and sucked on cinnamon candies called Atomic Fireballs. Doctors and others embraced the possibilities of rays that could travel through skin, letting them look inside the body. At shoe stores, customers could check whether their new shoes fit by putting their feet in an X-ray unit that showed their feet's bones inside the shoes.

DECREASED DEMAND FOR NUCLEAR POWER PLANTS

By the 1970s, 20 percent of the electricity in the United States came from nuclear power. It was very expensive to build nuclear power plants, however, and the enthusiasm for new plants began to wane. In 1979, there was an accident at the Three Mile Island nuclear plant in Pennsylvania. In 1986, the nuclear accident at Chernobyl in the Ukraine raised concern about the hazards of nuclear power. In addition, there was a growing stockpile of nuclear waste. Because of these problems, no new power plant has been ordered in the United States for more than thirty years.

Nuclear power may, however, come back into favor in the United States. As costs for coal, oil, and natural gas rise, people are looking to nuclear power as an option. Nuclear power may also offer an answer to an important environmental problem, global climate change. The carbon dioxide (CO_2) produced when coal or oil is burned may be causing the earth to warm. Carbon dioxide in the atmosphere forms a layer that lets in sunlight but traps heat over the earth. (It is similar to how the inside of a car heats up when its windows are closed on a summer day.) This warming may be contributing to the frequency of extreme weather events such as hurricanes, drought, and heat waves. Unlike coal and oil, nuclear fission produces no carbon dioxide, and increasing our use of nuclear power may offer a way to slow global warming.

FUEL FROM URANIUM

Most of the world's uranium comes from Canada, but it can be found almost anywhere. Much of the uranium in the United States was mined on Navajo reservations in the southwestern United States. Natural uranium has two different isotopes: uranium-235 and uranium-238. Uranium-235, the lighter isotope, can fission easily. Before it can be used as fuel, however, the uranium must be enriched. Power plants use uranium fuel that is only 3 to 7 percent enriched, while a nuclear bomb could be enriched to have 90 percent uranium-235.

In the enrichment process, the uranium ore goes through a mill that crushes and grinds the rock into a powder. Water and acid are added, then the mixture is dried to produce "yellow cake," a grainy yellow material. A ton (0.91 metric ton) of ore makes no more than four pounds (about 2 kilograms) of yellow cake. The yellow cake is combined with fluorine and heated until it becomes a gas.

The enrichment process separates the uranium-235 from the uranium-238. The remaining fuel, rich in uranium-235, is cooled and combined with other metals to form rods, pellets, or plates. These fuel pieces are then sealed in a metal container, usually in the shape of a long rod, and delivered to nuclear power plants.

THE NUCLEAR POWER PLANT

If you look at a nuclear power plant from a distance, you will see several large buildings and some dome-shaped ones. You might see a cloud of water vapor rising from a cooling tower. And you will see electric power wires, which deliver the electricity leaving the plant.

Inside each dome is a nuclear core, where long metal rods filled with uranium fuel pellets are bundled

This powdery yellow cake is a form of uranium oxide. The yellow cake seen here will be converted to a gas at an enrichment plant. After that, the gas can be processed into nuclear fuel or nuclear weapons.

together. Unlike in a nuclear bomb, however, where the chain reaction occurs in a split second, nuclear power plants are designed to release the energy stored in uranium atoms through a slow, controlled process. Therefore, among the fuel rods are control rods, which don't contain any uranium. The control rods are usually made of cadmium or boron, and their job is to absorb neutrons so the fission reactions will not take place too quickly. If the neutrons start to get out of control, the control rods automatically kick in to slow the reactions.

25
27
39
40
43

Inside the core, uranium throws off enough heat to boil the water that passes through it in tubes. The water turns to steam and is sent to the turbine hall.

In the large buildings, huge machines called turbines are turned by steam. This is the part of the plant that makes electricity. All electric plants have turbines, whether they are powered by nuclear fission, the burning of coal, wind, or even a river.

Turbines are based on a discovery made in 1831 by a scientist named Michael Faraday. Faraday discovered that electric current runs through a copper wire when the wire is moved near a magnetic field. To understand how electricity moves in a wire, imagine a row of seats in a theater. Perhaps someone in the aisle seat leans over and takes up the whole armrest that is supposed to be shared with the next person. The person in the second seat, feeling crowded, might lean over and take the other armrest, causing the next person to feel crowded and shift. This shift would move person by person down the row, until everyone is leaning in one direction.

This image shows the open core of a nuclear power plant in Vernon, Vermont. When the plant is operating, the core will be filled with rods of nuclear fuel. Nuclear reactors also have a core cooling system that uses water, sodium, or sodium salts as a coolant. The nuclear core becomes contaminated with use, so workers must wear protective clothing when performing any maintenance on the core.

Three Mile Island

If steam pressure builds up too much in a nuclear reactor, a valve will open to release the pressure. At Three Mile Island in Pennsylvania, in March 1979, a safety valve opened to release steam pressure, just as it should have. However, when the pressure went down again, the valve did not close. Steam continued to pour out through the valve until the reactor ran out of water. Without water, the reactor overheated and was damaged. Very little radioactivity was released, but the damaged parts of the plant had to be dismantled, disposed of, and handled as hazardous waste.

President and Mrs. Carter visit the control room of the Three Mile Island nuclear power plant after the accident in 1979. Before he entered political life, President Carter studied nuclear physics and worked on nuclear submarines.

A wire is made up of individual atoms, and each atom has electrons. If a magnet moves near the wire, the electrons in the atoms of the wire will shift. Each atom affects the next, and there is a shift of electrons down the line, like the people in the theater. The atoms do not actually move. Their electrons do, all in a split second. This shift can be measured as electric current.

Turn the magnet around and all the electrons shift back in the opposite direction.

If you could flip a magnet back and forth very fast near the wire, the electrons in the wire's atoms would shift back and forth very fast. This is exactly what a turbine does. Huge magnets are set around the edge of a wheel, with half of them facing in a positive direction and half in a negative direction. Then the wheel is spun so that the magnets pass by a coil of wire one after the other: positive, negative, positive. The electrons in the wire shift back and forth as the magnets pass, creating what is called alternating current.

To turn a huge magnet wheel very fast, you need a source of power. Nuclear, coal, and oil plants turn a turbine by boiling water into steam. As water boils, it expands and the burst of steam against fan blades spins the turbine. Once the steam has gone through the turbine, its job is done. Some plants just release the steam through cooling towers; others cool it back into water and reuse it in a cycle.

SPENT FUEL

After about eighteen months, much of the uranium in a fuel rod has fissioned and radioactive fission products have built up in the fuel. These used-up rods are called "spent" fuel. Not only are the spent fuel rods radioactive, but they also emit a tremendous amount of heat. Power plants store the spent fuel rods in large tanks of water for ten to twenty years while they cool and become less radioactive. Water shields the radiation so it does not leave the plant, and the water absorbs heat. Once the spent fuel has cooled, it is time to transport it to a permanent disposal place.

In every stage of uranium's production—at the mine, the mill, the enrichment plant, and finally the nuclear reactor—radioactive waste is produced. We do not know yet how long a power plant itself will "live," but we do

Fission Products

When the Curies detected radioactive rays coming from the uranium in their lab, they named the rays after the first three letters of the Greek alphabet. Alpha rays have a positive electrical charge. Beta rays have a negative charge. Gamma rays, which are high-energy photons, have no electrical charge. We know now that Alpha rays are really particles made up of two neutrons and two protons. Beta rays are made up of high-speed electrons. Alpha rays are stopped by paper. Beta rays can penetrate aluminum thinner than about an eighth of an inch (about 0.3 cm). Gamma rays can be stopped with a lead shield. All of these rays can be shielded with water, as long as it is deep enough to absorb the radioactivity. When spent fuel rods are transported from the reactor core to the storage pools, they travel along water-filled canals that keep them from emitting radiation into the air.

Spent fuel rods are stored in deep water in a storage tank. Above, some rods have already been placed around the perimeter of the storage rack. Both the concrete storage tank and the water bath shield the rods so radioactivity cannot escape. The water gradually cools the fuel rods. Even though they are no longer able to make electricity, the fuel rods will continue to produce heat for years.

know that at the end of its life, parts of the power plant will be radioactive and will be in need of proper disposal. Although the United States has been making nuclear power for over fifty years, a great deal of spent fuel and other nuclear waste is still in temporary storage awaiting final disposal. Much of the spent fuel is still at the power plants. Some waste is in tanks of water, and some has been moved to concrete casks.

4 TYPES OF WASTE

Low-level waste is stored in a trench at the Hanford site in Washington State.

Nuclear waste can take many forms, from the crushed rock at a uranium mine to small metal pellets of fuel, from liquids to glass blocks. Even shoes, tools, and medical syringes can contain radioactive waste.

A nuclear accident creates both waste and contamination. A nuclear submarine could sink or crash, releasing waste. A plant like Chernobyl could explode, spewing radioactive waste for hundreds of miles. The waste produced from each of these accidents would have to be contained, transported, and stored.

Radioactive waste is not allowed in a regular landfill. The United States government divides waste into three categories: uranium-mining waste, low-level waste, and high-level waste.

WASTE FROM URANIUM MINING AND MILLING

Uranium tailings are the leftovers from the mining and processing of natural ore to extract uranium. Because only a few pounds of yellow cake can be taken from a ton of uranium ore, waste from mining and milling makes up the giant's share of the world's nuclear waste. In the United States alone, there are dozens of mine sites. As much as five billion cubic feet (142 million cubic meters) of tailings were piled into hills and left at these mine sites for many years. The waste piles still contained uranium, and they emitted dangerous radon gas. Wind and rain washed the radioactive dust from waste piles into streams and rivers.

Some of the mining waste in the United States actually came from African ore. Before World War II, a man named Edward Sengier, who managed mines in Belgian Congo, shipped ore to the United States. The ore sat in a warehouse in Staten Island, New York, until 1942, when an engineer working for the United States government called Sengier, asking how to get African uranium. Imagine the engineer's surprise when Sengier told him he would find African ore in a Staten Island warehouse. This ore fueled the first nuclear chain reaction in Chicago, Illinois, and it supplied the plutonium and uranium for the bombs dropped on Hiroshima and Nagasaki. It also left a trail of radioactive waste disposal sites across the United States, which, to this day, still need to be cleaned up.

In 1978, to tackle the problem of mill waste, the United States Congress passed a law called the Uranium Mill Tailings Radiation Control Act. Because of this act, the United States Nuclear Regulatory Commission now requires uranium milling operations to be located far from rivers and places where many people live. The Nuclear Regulatory Commission encourages companies to bury the waste and grow plant cover over it to reduce wind and water erosion. In the meantime, the federal government has been working to clean up mill tailings sites.

LOW-LEVEL WASTE

Low-level waste is anything that has become contaminated with radioactive material or has itself become radioactive. Workers' shoe covers and clothing can become contaminated, for example, as can cleaning tools such as mops and rags. Not all low-level waste comes from nuclear power plants. Hospitals and laboratories

The steel drums stored at this waste site contain low-level waste. The U.S. Nuclear Regulatory Commission separates low-level waste into three different classes: A, B, or C. Class A waste (shown in this photograph) has the lowest levels of radioactivity.

produce low-level radioactive materials such as swabs, needles, and even bodies of lab animals. Before it can be shipped for disposal, low-level waste must be placed in special containers approved by the United States Department of Transportation. It is buried in a low-level waste landfill, which is watched over by state government inspectors. Only three low-level waste disposal sites currently exist in the United States. These facilities have been designed so that no radioactive material can escape into the air or underground water.

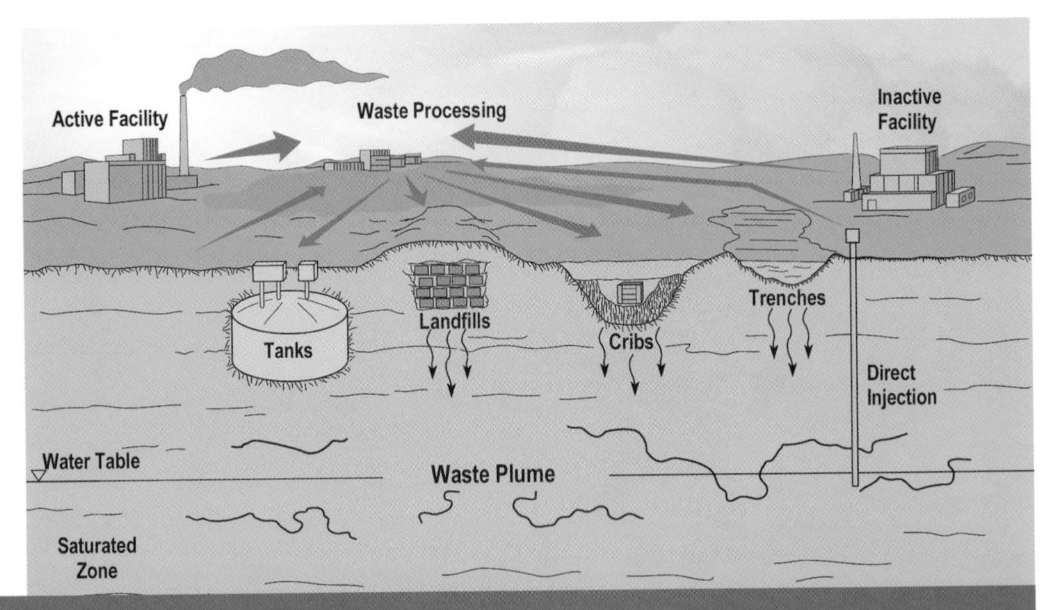

For more than fifty years, before the health risks of radioactivity were better understood, the United States built a large network of laboratories and factories to research, build, and test nuclear weapons and nuclear fuel. Thousands of locations now have improperly buried waste that has spread to underground soil and water. The U.S. Department of Energy estimates that it will take about $300 billion and seventy years to clean up all of these locations.

Many of the old low-level waste sites, however, have leaked waste into the nearby environment. Maxey Flats, Kentucky, was one such low-level waste landfill. In 1970, seven years after it opened, inspectors found that plutonium had traveled underground away from the site. This was surprising, since plutonium does not penetrate soil well. Upon further investigation, geologists discovered that the waste trenches had been dug over sand deposits; plutonium travels easily through sand. The

Maxey Flats landfill had to be closed, but the waste was left there. By 1980, radioactive waste had traveled about 2 miles (over 5 km) away from the Maxey Flats site to drinking water wells and streams. The site has been "contained" to prevent further spread, but cleanup is still going on. Because of this occurrence, much more care is now taken in choosing locations for radioactive waste disposal.

Decommissioned Power Plants

At a certain point, a nuclear power plant will begin to wear out and it will need to be closed and taken apart. Over the life of the plant, buildings and equipment become radioactive. Decommissioned plants are classified as low-level waste. The reactor core, even with the fuel rods removed, could be so contaminated it would be hazardous work to take it apart. In 1989, an entire reactor core was sent from Pennsylvania by barge to Port Madison, Washington, to be buried as low-level waste. It traveled 8,000 miles (about 13,000 km) down the Atlantic coast, through the Panama Canal, and then up the Pacific coast.

Transuranic Waste

"Transuranic" literally means "beyond uranium," and transuranic waste is a special type of low-level waste. It is made up of elements, such as plutonium, that are heavier than uranium because they have more protons and neutrons. Plutonium emits mostly alpha rays, which are not very strong. A sheet of paper or an inch (about 2.5 cm) of air will stop an alpha ray. While transuranic waste is not as hazardous as other forms of radiation, it lasts a very long time. For this reason, the United States government has decided to store it in a different, isolated location. Transuranic waste has been separated from other low-level waste since about 1999.

HIGH-LEVEL WASTE

High-level radioactive waste includes spent reactor fuel rods and waste materials left over when spent fuel is reprocessed. The military also produces high-level waste when building and recycling nuclear weapons. Until it can be shipped to a permanent location, spent fuel is kept at nuclear power plants in great vats of water or concrete casks. For many years, the United States reprocessed or recycled spent fuel, removing the uranium so it could be used again. The United States no longer reprocesses spent fuel, but there is a great deal of waste left from past reprocessing, some of it liquid and some of it molded in glass blocks. The military still operates reprocessing plants for waste from weapons. High-level waste must be handled, transported, and stored very carefully.

In 2003, as part of the major cleanup effort at the contaminated Hanford nuclear reservation in Richland, Washington, these buildings were demolished. The facility was used to make plutonium from uranium for one of the atomic bombs used in World War II.

CLEANING UP MISTAKES FROM THE PAST

On first glance, Richland, Washington, may look like an ordinary American town. It is surrounded by sagebrush desert. The Cascade Mountains are visible in the distance, and the Columbia River flows nearby. The town has grocery stores and other businesses. Parents go to work. Children go to school. Look more closely, and you might notice a store or business with the word "Atomic" in its name. Most of the houses are one of several styles, built by the government in the 1940s. And you might notice that Richland's high school mascot is a mushroom cloud, the kind of cloud made by an atomic bomb.

What you will not see is that under the ground, not far from Richland, there are close to 200 tanks of nuclear waste, many of them leaking. You also will not see the radioactive contamination seeping into the Columbia River.

Before 1943, few people lived where Richland now stands. In 1943, however, the population boomed from about 1,500 residents to more than 51,000 when the United States government built a laboratory called Hanford Nuclear Reservation. Three nuclear reactors, three chemical plants, and underground storage tanks were built, as well as thousands of houses for the employees and their families. People from across the country moved to work on a top-secret project. It was so secret that many of the reservation's workers did not even know that what they were doing was making the fuel for an atomic bomb. That is why Richland's sports teams still boast the atomic mascot.

But years ago, people did not realize that in order to prevent harming people and the environment, nuclear waste must be properly handled, stored, moved, and buried. Because people did not know how to handle nuclear waste, the area now has underground

The Canister Storage Building *(above)* at the Hanford Nuclear Reservation in Richland, Washington, will store 2,300 tons (2,086 metric tons) of high-level radioactive waste. The circles on the floor are openings to tubes where dried spent fuel will be inserted and placed in reinforced concrete underground vaults. A cooling system will keep the waste from overheating.

radioactive pollution, and the Hanford site is currently going through one of the most massive cleanups in United States history.

Many of the other laboratories that were created during World War II to make the first atomic bombs are also still holding radioactive waste, and in many of these storage sites, the waste has leaked into the soil or water. Local citizens, journalists, and lawmakers are working to make sure the government cleans up these waste sites.

5 WHAT TO DO WITH THE WASTE

Located in the Nevada desert, Yucca Mountain will be the final burial place for the nation's high-level radioactive waste.

In the 1950s, when the United States government was encouraging private companies to build nuclear power plants, it promised to come up with a disposal plan for the spent fuel. Thirty years later, there was still no plan and spent fuel rods were piling up at power plants, either in water pools or in cement casks after they had cooled.

The 1982 Nuclear Waste Policy Act

The accident in 1979 at the Three Mile Island power plant in Pennsylvania created a large amount of nuclear waste, and it reminded the public and Congress of the hazards of nuclear power. In response, Congress passed a law on the disposal of spent fuel and other radioactive waste. The 1982 Nuclear Waste Policy Act stated that the federal government must find a place to permanently dispose of high-level radioactive waste and spent nuclear fuel. The companies operating the waste-producing power plants would pay for the disposal of the fuel rods. In addition, the law said that the public would have a say in planning the site.

The 1982 law also addressed the fact that waste had been improperly buried or dumped in many locations throughout the United States. The Department of Energy would have to clean up 253 massive underground tanks, containing approximately a hundred million gallons (379 million liters) of liquid and sludge high-level radioactive waste in Washington, Idaho, South Carolina, and New York. Furthermore, the Nuclear Waste Policy Act required the Department of Energy to clean up the waste that leaked into the environment from these storage tanks and dispose of it in a cavern deep under the ground.

Looking for Solutions

The long life of radioactive waste poses challenges to the people who are looking to find a safe location to store it. It can take 10,000 years for some radioactive waste to decay into something that is no longer dangerous. How long is 10,000 years? Ten thousand years ago, people spoke different languages from those we speak today. Governments and countries have changed. Climates have also changed dramatically, becoming wetter or warmer. Earthquakes and volcanoes have altered landscapes.

Scientists and engineers are tackling the task of designing a storage facility that can withstand the changes of time. Many of these people work at the United States government's Exploratory Studies Facility. For two decades, the Office of Civilian Radioactive Waste Management has conducted scientific and engineering investigations at Yucca Mountain, Nevada, to determine whether it would work as a nuclear waste repository. No one lives on Yucca Mountain. The closest town is about 14 miles (over 23 km) away. The Yucca Mountain area is surrounded by U.S. government-controlled land and has been used in the past for underground nuclear tests.

Scientists responsible for nuclear waste disposal decided that deep burial would be the best idea. Congress agreed and wrote deep burial right into the law. The scientists selected Yucca Mountain, and the United States government is currently building tunnels hundreds of stories deep into the mountain to hold 77,000 tons (69,800 metric tons) of fuel rods and other high-level waste. Waste containers will be made of a

This tunnel reaches 5 miles (8 km) down into Yucca Mountain. In underground laboratories, engineers and scientists test the mountain's rock, heating and freezing it, and measuring its strength to verify that this will be a safe place to store radioactive waste. The United States government plans to dig 50 miles (about 80 km) of tunnels to store as much as 77,000 tons (69,800 metric tons) of high-level waste in sealed containers. The site will be the world's largest storehouse of nuclear waste.

metal alloy, which should last for 10,000 years if they stay dry. One reason Yucca Mountain was selected for the deep tunnels is that it is located in a desert, and since there is little rain, water is less likely to reach the storage containers. In addition, the area's water table, many miles underground, is one of the lowest water tables in the world.

There have been many legal difficulties for the Yucca Mountain plan. Citizens, communities, and even the state government have resisted the creation of a nuclear waste site in Nevada. People have protested, spoken at public meetings, and filed lawsuits. The Yucca Mountain plan took a big step forward in 2002, however, when the president of the United States, George W. Bush, signed a bill to overrule the objections of the state of Nevada.

There have also been technical challenges in building the site. As they have looked for a place to store radioactive waste, scientists and engineers have had to consider many questions:

How do you move waste to the disposal site?

Spent nuclear fuel, before being transported in the United States, must first be sealed in special containers called casks, which shield the waste so radiation cannot escape. In some casks, all water and air are removed, leaving a vacuum. The canister is filled with a gas and sealed. Some casks are welded, and some are bolted shut.

Thousands of truckloads and railcars of nuclear waste have traveled across the United States. There have never been any accidents, yet transportation of spent fuel worries the public. Hundreds of cities and counties have tried to ban shipment of spent fuel on their roads. Some laws have allowed only nighttime passage and have required armed guards to accompany the trucks.

Will heat from the radioactive material break down the storage containers?

After sitting in temporary storage for decades, transuranic waste is being trucked in casks from Idaho to a permanent facility in New Mexico.

Storage containers and chambers must be able to handle the damaging power of radioactive heat. Because the chamber is sealed, even slightly hot radioactive material could build up over time. This heat could hasten corrosion of the containers or build up enough pressure to burst container walls if they are not strong enough. Hopefully they will withstand these pressures.

Will water that seeps in from the earth cause containers to corrode or carry radioactivity into groundwater?

Although Yucca Mountain is located in a desert now, and it has one of the lowest water tables in the world,

These casks contain spent fuel from the Savannah River Site research center in South Carolina. To prevent radiation from escaping, spent fuel rods are transferred from stainless steel casks to the aluminum containers shown above under seventeen feet (five meters) of water. The containers will eventually be sent to Yucca Mountain.

the climate could change and become wet in the future. Water can corrode waste containers, dissolve radioactive waste, and carry toxic pollution underground toward drinking water wells.

Could an earthquake break open a storage cavern?

Yucca Mountain is made of hard volcanic rock. While this rock makes a good water barrier, geologists point out that where there is volcanic rock, there is also potential for earthquakes. A 1992 earthquake shook the Yucca Mountain project field office, causing damage. Future earthquakes could create cracks in the rock, allowing rainfall to seep down into storage caverns. To compensate for this, waste is sealed in waterproof containers.

These dry casks at a nuclear waste storage facility in Prairie Island, Minnesota, hold spent fuel until it can be shipped to a permanent storage vault at Yucca Mountain.

How do you completely seal off the containers from human contact but then make sure the containers are not leaking?

One of the difficulties of buried waste is keeping track of its condition. Radioactive pollution could travel miles in the groundwater before anyone discovers it. If workers reopen the burial ground to stop the leakage into groundwater, they would be entering unknown hazardous conditions, and they might accidentally release pollution into the air. Some countries, such as Great Britain, have opted for storage above the ground, so if any casks crack, they would see liquid waste leaking out.

How do you tell future inhabitants of the earth that this waste is dangerous?

Some radioactive waste will continue to emit radiation for thousands of years. Language and culture experts were asked to help figure out how to label the radioactive waste so that future populations will know it is dangerous, even though they may no longer speak any of the languages spoken in the world today. The experts suggested that menacing stone spikes and pictures of horrified and sick faces be built at the opening to the Yucca Mountain tunnels.

If you have ever walked through an old cemetery and seen gravestones worn smooth by wind and rain, you know that markers do not last forever. Not only would the waste site's markers or labels need to look scary, they would need to stay that way for thousands of years. Engineers are searching for materials that are strong enough to withstand the wear of weather and time.

Learning from Ancient Reactors

Can Yucca Mountain keep nuclear waste safely out of reach for thousands of years? Natural reactors found in Africa have given scientists and engineers hope that this is possible by offering a full-scale model to study. Two billion years ago, deep underground in Oklo, Gabon,

Africa, water mixed with uranium in just the right amounts. The uranium began a nuclear chain reaction, emitting heat and radiation. The heat boiled away the water, and with the water gone, the nuclear reaction stopped. After the rock cooled and more water seeped in, the chain reaction began again. The nuclear reactions started and stopped and started again over a million years until the special form of uranium was used up.

In 1972, a French physicist named Francis Perrin discovered the evidence of these ancient nuclear reactors. He figured the reactions must have occurred there a billion years ago because there was so little left of that particular form of uranium, uranium-235, which is capable of nuclear reaction. In addition, he found some of the waste products that are created in nuclear reactors. The most interesting thing about the natural nuclear reactors at Oklo is that the radioactive waste products seem to have stayed in place, underground, for all that time. Scientists are excited about the discovery at Oklo, Gabon, because it may help clarify how to safely store radioactive waste deep under the ground, protecting present and future generations.

Difficult Energy Choices

The United States gets electric power primarily from coal, oil, natural gas, and nuclear power, and each of these energy sources has its own health hazards for workers. Like uranium miners, coal miners die when mines cave in. Miners and power plant workers sometimes suffer from lung diseases and cancer caused by the dust and gas they breathe on the job, though improved equipment now helps protect their lungs. Drilling and refining oil and natural gas can also be dangerous. Workers use heavy equipment and often work in extreme outdoor environments. Gas can explode, or hazardous gas can escape from a well. There are regulations to protect these workers, but accidents can happen.

The fuels we use for energy can harm the environment and the health of people living far from the power plants. Coal and oil can contaminate drinking water, and they can produce air pollution that causes acid rain and harmful smog hundreds of miles away. There is also the problem of global climate change. Unburned natural gas and the carbon dioxide produced when coal and oil are burned may contribute to global warming. In time, this warming could create changes to our climate that could have a devastating effect on the environment and human health. If we had not been using nuclear power for the past sixty years, the world's non-nuclear power plants would have put an extra 1,600 million tons (1,456 million metric tons) of carbon dioxide into the atmosphere every year.

What you can do to help the environment:

1) All forms of electric power create pollution or in some way change the environment. Find ways you can use less electricity. For example, could you take more care to turn off lights and computers when you are not using them? Could you encourage friends and family to buy more efficient light bulbs?
2) Learn about forms of energy that produce less pollution, such as wind and solar power.
3) Learn about energy-efficient cars and appliances.
4) Find out how and where your electricity is made. Ask a teacher or parent the name of your community's electric company and look at the company's Web site. Does your electricity come mostly from coal or oil? Is it nuclear? Some companies let you sign up to use more solar and wind power.
5) Learn more about nuclear energy. To help teach the public, many nuclear power plants offer public tours.

GLOSSARY

atom The smallest particle of an element. Atoms are made up of protons, electrons, and neutrons.

control rod A rod used to absorb neutrons in a nuclear reactor, slowing the fission chain reaction.

decommission To take a power plant out of service and close it down.

electron A negatively charged particle.

element A substance made up of a single type of atom.

fission The splitting of an atom's nucleus, which releases heat and energy.

fuel rod A sealed rod filled with uranium fuel, used to power a nuclear reactor.

half-life The time it takes for a radioactive substance to give off half its radiation.

isotope An element with the same number of protons but a different numbers of neutrons.

molecule A substance made up of two or more types of atoms.

neutron A particle in the nucleus of an atom that has no electric charge.

photon A particle having energy and momentum but no mass or electric charge.

pitchblende A radioactive rock that contains radium and uranium.

plutonium A radioactive element created in nuclear reactors.

proton A particle in the nucleus of an atom that has a positive electric charge.

radiation Energy that is transferred by rays as atoms or molecules undergo change.

radioactivity The spontaneous release of neutrons, alpha or beta particles, or gamma rays from an atom with an unstable nucleus.

reprocessing Recycling of fuel rods to save or reuse remaining uranium.

spent fuel Fuel that has used up some of its uranium and built up fission products.

transuranic Relating to elements having higher numbers (and therefore heavier) than uranium.

turbine An engine that extracts energy from a flow of fluid or steam.

uranium A natural, radioactive element found in pitchblende.

uranium-235 An isotope of uranium that undergoes fission and is used as fuel for nuclear reactors and nuclear weapons.

uranium-238 An isotope of uranium used to make plutonium.

water table The level below which the ground is saturated with water.

For More Information

Department of Energy
Energy Information Administration
1000 Independence Avenue SW
Washington, DC 20585
(202) 586-8800
Web site: http://www.eia.doe.gov/kids/history/timelines/nuclear.html

Nuclear Energy Institute
1776 I Street NW, Suite 400
Washington, DC 20006-3708
(202) 739-8000
Web site: http://www.nei.org/scienceclub/index.html

Nuclear Information and Resource Service
1424 16th Street NW, #404
Washington, DC 20036
(202) 328-0002
Web site: http://www.nirs.org

U.S. Nuclear Regulatory Commission
Office of Public Affairs (OPA)
Washington, DC 20555
(800) 368-5642
Web site: http://www.nrc.gov/reading-rm/basic-ref/students.html

WEB SITES

Due to the changing nature of Internet links, the Rosen Publishing Group, Inc., has developed an online list of Web sites related to the subject of this book. This site is updated regularly. Please use this link to access the list:

http://www.rosenlinks.com/eet/rawa

FOR FURTHER READING

Galperin, Anne L. *Nuclear Energy/Nuclear Waste* (Earth at Risk). New York, NY: Chelsea House Publishers, 1991.

Gore, Gordon. *Experimenting with Energy* (Experimenting With . . .). Ontario, Canada: Trifolium Books, 2000.

Hare, Tony. *Nuclear Waste Disposal* (Save Our Earth). London, England: Aladdin Books Ltd., 1990.

Scarborough, Kate. *Nuclear Waste* (Our Planet in Peril). Mankato, MN: Bridgestone Books, 2002.

Wilson, P. D. *The Nuclear Fuel Cycle: From Ore to Waste* (Oxford Science Publications). Oxford, England: Oxford University Press, 1996.

Yanda, Bill. *Rads, Ergs, and Cheeseburgers: The Kids' Guide to Energy and the Environment*. Santa Fe, New Mexico: J. Muir Publications, 1991.

BIBLIOGRAPHY

Bartlett, Donald L., and James B. Steele. *Forevermore: Nuclear Waste in America*. New York, NY: W. W. Norton and Company, 1985.

BBC News. Reports on the Chernobyl Disaster. April 22, 2000; September 3, 2004; and November 20, 2004. Retrieved June/July 2005 (http://www.bbc.co.uk/dna/h2g2/A2922103).

Blatt, Harvey. *America's Environmental Report Card: Are We Making the Grade?* Cambridge, MA: MIT Press, 2005.

CEKERT Nuclear Energy Technology Centre. "Virtual Nuclear Reactor Tour." Retrieved June/July 2005 (http://www.cekert.kth.se/nuclear_power/virtual/nobel_showcase/reactor_site.html).

Department of Energy, Energy Information Administration. "Energy Kid's Page." October 2005. Retrieved June/July 2005 (http://www.eia.doe.gov/kids/history/timelines/nuclear.html).

Department of Energy, Office of Civilian Radioactive Waste Management. Retrieved June/July 2005 (http://www.ocrwm.doe.gov/factsheets/doeymp0010.shtml).

Department of Energy, Office of Nuclear Energy, Science and Technology. "Splitting Atoms—An Electrifying Experience." Video. Retrieved June/July 2005 (http://www.nuclear.gov/home/public1.html).

Domenici, Pete, *A Brighter Tomorrow: Fulfilling the Promise of Nuclear Energy*. Lanham, MD: Rowman and Littlefield, 2004.

Easton, Thomas A., ed. *Taking Sides: Clashing Views on Controversial Environmental Issues*. Dubuque, IA: McGraw-Hill, 2005.

Frame, Paul W. Oak Ridge Associated Universities. *Tales from the Atomic Age*. "Jááchymov: Cradle of the Atomic Age." Retrieved June/July 2005 (http://www.orau.org/ptp/articlesstories/jachymov.htm).

Green, Jen. *Energy Crisis*. North Mankato, MN: Chrysalis Education, 2003.

Lawrence Berkeley National Laboratory. "ABCs of Nuclear Science." April 4, 2005. Retrieved June/July 2005 (http://www.lbl.gov/abc/Basic.html#Radioactivity).

Lipper, Ilan, and Jon Stone. University of Michigan, "Energy and Society." Retrieved June/July 2005 (http://www.umich.edu/~gs265/society/nuclear.htm).

McCarthy, John. Stanford University. "Frequently Asked Questions about Nuclear Energy." October 1995. Retrieved June/July 2005 (http://www_formal.stanford.edu/jmc/progress/nuclear_faq.html).

Morton, Sheryl. U.S. Department of Energy National Spent Nuclear Fuel Program. "What Is Spent Nuclear Fuel?" June 2000. Retrieved June/July 2005 (http://nsnfp.inel.gov/whatis.asp).

Nuclear Energy Institute. "NEI Science Club." August 2000. http://www.nei.org/scienceclub/index.html.

Nuclear Weapons Archive. "Trinity, 16 July 1945." Retrieved June/July 2005 (http://nuclearweaponarchive.org/Usa/Tests/Trinity.html).

Seattle Times. "Fifty Years from Trinity." 1995. Retrieved June 2005 (http://seattletimes.nwsource.com/trinity).

Shapiro, Fred C. *Radwaste: A Reporter's Investigation of a Growing Nuclear Menace*. New York, NY: Random House, 1981.

Shrader-Frechette, K. S. *Burying Uncertainty: Risk and the Case Against Geological Disposal of Nuclear Waste*. Berkeley, CA: University of California Press, 1993.

Stoyles, Pennie, Peter Pentland, and David Demant. *Nuclear Energy*. North Mankato, MN: Smart Apple Media, 2004.

U.S. Department of Energy. "Uranium Stewardship Activities, Uranium Facts." Retrieved June/July 2005 (http://www.ne.doe.gov/uranium/facts.html).

U.S. Nuclear Regulatory Commission. Retrieved June/July 2005 (http://www.nrc.gov/).

Waltham, Chris. University of British Columbia, Canada. "Cloud Chambers: An Interactive Look at Radiation." Retrieved June/July 2005 (http://www.physics.ubc.ca/~outreach/phys420/p420_97/chris/p1.htm).

World Nuclear Association. "Glossary of Nuclear Terms." December 2002. Retrieved June/July 2005 (http://www.world-nuclear.org/info/inf51.htm).

INDEX

ABOUT THE AUTHOR

D. D. Kelly is an environmental engineer and writer. She has written about science and the environment for many magazines.

PHOTO CREDITS

Cover, pp. 33, 34, 37 © Roger Ressmeyer/Corbis; pp. 1, 50, 51 © Peter Essick/Aurora Photo; p. 4 © Getty Images, Inc.; pp. 7, 13, 17, 19, 28, 43, 49 © AP/Wide World Photos; pp. 8, 23 © Bettmann/Corbis; p. 10 © Scott Camazine/Photo Researchers, Inc.; pp. 11, 22 © Corbis; p. 15 © Astrid and Hanns-Frieder Michler/Photo Researchers, Inc.; p. 20 © Hank Morgan/Photo Researchers, Inc.; p. 27 © Paolo Koch/Photo Researchers, Inc.; p. 30 © Time Life Pictures/Getty Images, Inc.; p. 38 © U.S. Department of Energy Genomics: GTL Program; p. 41 © AP/Wide World Photos/Department of Energy; p. 44 © Dan Lamont/Corbis; p. 47 © AFP/Getty Images, Inc.

Designer: Thomas Forget; Editor: Liz Gavril; Photo Researcher: Hillary Arnold